GOODS OF THE MIND, LLC

Competitive Mathematics Series

for

Gifted Students in Grades 3 and 4

PRACTICE COMBINATORICS

Cleo Borac, M. Sc.
Silviu Borac, Ph. D.

This edition published in 2014 in the United States of America.

Editing and proofreading: David Borac, M.Mus.
Technical support: Andrei T. Borac, B.A., PBK

Send all inquiries to:

Goods of the Mind, LLC
1138 Grand Teton Dr.
Pacifica
CA, 94044

Competitive Mathematics Series for Gifted Students
Level II (Grades 3 and 4)
Practice Combinatorics
2nd edition

Contents

FOREWORD

The goal of these booklets is to provide a problem solving training ground starting from the earliest years of a student's mathematical development.

In our experience, we have found that teaching how to solve problems should focus not only on finding correct answers but also on finding better solution strategies. While the correct answer to a problem can typically be obtained in several different ways, not all these ways are equally useful for learning how to solve problems.

The most basic strategy is *brute force*. For example, if a problem asks for the number of ways Lila and Dina can sit on a bench, it is easy to write down all the possibilities: Dina, Lila and Lila, Dina. We arrive at this solution by performing all the possible actions allowed by the problem, leaving nothing to the imagination. For this last reason, this approach is called brute force.

Obviously, if we had to figure out the number of ways 30 people could stand in a line, then brute force would not be as practical, as it would take a prohibitively long time to apply.

Using brute force to obtain the correct answer for a simpler problem is not necessarily a useful learning experience for solving a similar problem that is more complex. Moreover, solving problems in a quantitative manner, assuming that the student can transfer simple strategies to similar but more complex problems, is not an efficient way of learning problem solving.

From this simple example, we see that the goal of *practicing* problem solving is different from the goal of problem solving. While the goal of problem solving is to obtain a correct answer, the goal of practicing problem solving is to acquire the ability to develop strategies, generate ideas, and combine approaches that are powerful enough to solve the problem at hand as well as future similar problems.

While brute force is not a useless strategy, it is not a key that opens every door. Nevertheless, there are problems where brute force can be a useful tool. For instance, brute force can be used as a first step in solving a complex problem: a smaller scale example can be approached using brute force to help the problem solver understand the mechanics of the problem and generate ideas for solving the larger case.

All too often, we encounter students who can quickly solve simple problems by applying brute force and who become frustrated when the solving methods they have been employing successfully for years become inefficient once problems increase in complexity. Often, neither the student nor the parent has a clear understanding of why the student has stagnated at a certain level. When the only arrows in the quiver are guess-and-check and brute force, the ability to take down larger game is limited.

Our series of books aims to address this tendency to continue on the beaten path - which usually generates so much praise for the gifted student in the early years of schooling - by offering a challenging set of questions meant to build up an understanding of the problem solving process. Solving problems should never be easy! To be useful, to represent actual training, problem solving should be challenging. There should always be a sense of difficulty, otherwise there is no elation upon finding the solution.

Indeed, practicing problem solving is important and useful only as a means of learning how to develop better strategies. We must constantly learn and invent new strategies while questioning the limitations of the strategies we are using. Obtaining the correct answer is only the natural outcome of having applied a strategy that worked for a particular problem in the time available to solve it. Obtaining the wrong answer is not necessarily a bad outcome; it provides insight into the fallacies of the method used or into the errors of execution that may have occured. As long as students manifest an interest in figuring out strategies, the process of problem solving should be rewarding in itself.

Sitting and thinking in a focused manner is difficult to train, particularly since the modern lifestyle is not conducive to adopting open-ended activities. This is why we would like to encourage parents to pull back from a quantitative approach to mathematical education based on repetition, number of completed pages, and the number of correct answers. Instead, open up the time boundaries that are dedicated to math, adopt math as a game played in the family, initiate a math dialogue, and let the student take his or her time to think up clever solutions.

Figuring out strategies is much more of a game than the mechanical repetition of stepwise problem solving recipes that textbooks so profusely provide, in order to "make math easy." Mathematics is not meant to be easy; it is meant to be interesting.

Solving a problem in different ways is a good way of comparing the merits of each method - another reason for not making the correct answer the primary goal of the activity. Which method is more labor intensive, takes more time or is more prone to execution errors? These are questions that must be part of the problem solving process.

In the end, it is not the quantity of problems solved, the level of theory absorbed, or the number of solutions offered in ready-made form by so many courses and camps, but the willingness to ask questions, understand and explore limitations, and derive new information from scratch, that are the cornerstones of a sound training for problem solvers.

These booklets are not a complete guide to the problem solving universe, but they are meant to help parents and educators work in the direction that, aside from being the most efficient, is the more interesting and rewarding one.

The series is designed for mathematically gifted students. Each book addresses an age range as some students will be ready for this content earlier, others later. If a topic seems too difficult, simply try it again in a couple of months.

SETS

Sets are well defined collections of objects. Some examples:

- The set of frying pans in my kitchen.
- The set of whales on this planet.
- The set of integer numbers.

When an object belongs to a set, we say it is an *element* of that set.

We can describe a set by *enumerating* (i.e. listing) its elements, as in this set of digits:

$$\{0, 1, 2, 3, 4, 5, 6, 7, 8, 9\}$$

or we can describe sets by drawing a simple picture of them, known as a *Venn diagram*:

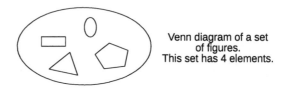

Venn diagram of a set
of figures.
This set has 4 elements.

Experiment

Think up sets using objects from your environment. For each of them:

1. Write up a definition in words.
2. Make a Venn diagram.
3. If possible, write up a description by enumeration.

Some sets are part of other sets. For example, the set of snakes is *included* in the set of reptiles:

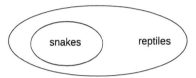

We call set of snakes a *subset* of the set of reptiles.

More examples of subsets:

- The set of manatees is a subset of the set of mammals.
- The set of 3 digit numbers is a subset of the set of integer numbers.
- The set of black cars is a subset of the set of cars.

Experiment

Think up sets that have subsets and make a Venn diagram for each example.

The Empty Set, also called the *null set*, is a set without any elements. There is only one such set. If you don't have a solution you can say "the set of solutions is empty." But first make sure this is true!

Sometimes, sets do not share any elements, such as the set of frying pans in my kitchen and the set of trees in my garden. We call such sets *disjoint*. Their representations as Venn diagrams might look like this:

Disjoint sets

Other times, sets share some of their elements. For example, the set people on a bus and the set of dog-owners may have elements in common if any of the people on the bus also own a dog. The Venn diagrams of these sets will then *overlap*:

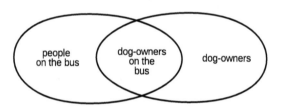

When we count the total number of elements in the sets we must make sure we do not count the dog-owners who are on the bus twice!

Do not use a calculator for any of the problems!

Exercise 1

Draw a Venn diagram that represents the following sets:

- the set of tennis balls at Stephan's club
- the set of Wilson tennis balls at Stephan's club
- the set of pink tennis balls at Stephan's club

Exercise 2

(i) Draw a Venn diagram that represents the following sets:

- Set W: the set of vegetables Max displayed in the store this morning
- Set X: the set of potatoes Max displayed in the store this morning
- Set Y: the set of purple potatoes Max displayed in the store this morning
- Set Z: the set of yellow flesh potatoes Max displayed in the store this morning

(ii) Which of the sets in your diagram are disjoint?

(A) set Z (yellow flesh potatoes) and set X (potatoes)

(B) set Y (purple potatoes) and set W (vegetables)

(C) set Y (purple potatoes) and set Z (yellow flesh potatoes)

(D) set W (vegetables) and set X (potatoes)

(iii) In your diagram, which sets are subsets of the set of vegetables?

Exercise 3

Dina went to a wilderness survival class. The girls and boys in the class went to gather mushrooms. Some of them were lucky and found mushrooms and some of them did not. Among the students who found mushrooms, some were girls and some were boys. Make a Venn diagram that illustrates the various sets and subsets in this story.

Now, fill in the blanks in the following paragraph:

The set of boys is a .... *of the set of students in the survival class. The set of girls is a* .... *of the set of students in the survival class. The set of boys and the set of girls are* .... *The set of girls and boys who found mushrooms is a* .... *of the set of students in the survival class.*

Exercise 4

Which of the following describes two subsets that must be disjoint? Check all that apply.

(A) At 1 PM, there were 18 cars in the parking lot at Alfonso's grocery. 4 cars were blue and 14 cars had tinted windows.

(B) At 1 PM, there were 18 cars in the parking lot at Alfonso's grocery. 5 cars had license plates that started with the letter Z and 3 cars had license plates that started with the letter Y.

(C) At 1 PM, there were 18 cars in the parking lot at Alfonso's grocery. 3 cars belonged to staff and 14 cars belonged to customers.

(D) At 1 PM, there were 18 cars in the parking lot at Alfonso's grocery. 8 cars were hatchbacks and 10 cars had roof racks.

Exercise 5

In a deck of playing cards, which of the following are disjoint subsets? Check all that apply.

(A) the picture cards and the diamonds

(B) the diamonds and the 4s

(C) the picture cards and the 4s

(D) the clubs and the diamonds

Exercise 6

Amira's class went on a field trip to the Museum of Art. Of the 24 students in the class, 15 used an audioguide and 18 took notes in their notepads. How many students used an audioguide and also took notes?

Exercise 7

Max, the baker, made 48 pieces of pastry this morning. Of these, 30 contain chocolate and 25 contain almonds. How many pieces of pastry contain both chocolate and almonds?

Exercise 8

Cornelia has 30 ducks on her farm. Every second day, 16 of them each lay an egg. The rest each lay an egg every third day. What is the largest number of eggs Cornelia may collect in one week?

Exercise 9

Stephan, the tennis coach, received 44 applications for a mock doubles tournament. In a doubles game, there are either 4 female players or 4 male players. Stephan must admit or reject each application. If each participant plays only once in the preliminary round, what is the largest number of applicants Stephan may have to reject?

Exercise 10

Patrick, the leader of the math club at Amira's school, asked the club members for their sizes, so he could order T-shirts with the club logo for them. The club members gave him their sizes in the form of a riddle: "All of us are size S, except for six of us. All of us are size M, except for seven of us. All of us are size L, except for nine of us." How many members are in the club have? How did Patrick figure out how many T-shirts of each size he should order?

Exercise 11

Consider the following Venn diagram:

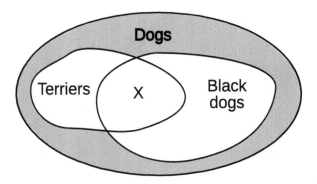

1. How would you describe the set X?
2. Make a list of the sets that are subsets of other sets. How many items are there?
3. How would you describe the shaded part of the set "Dogs?"

LAST DIGIT OF A SUM OR PRODUCT

Since there can be no carryover in the units place of an addition, it is always sufficient to add the last digits of all the terms to find the last digit of the sum. There following sum, for instance:

$$127 + 3459 = 3586$$

has the same last digit as the sum of the last digits of its terms:

$$7 + 9 = 16$$

Likewise, the last digit, the last digit of a multiplication is always the same as the last digit of the product of the last digits of the factors. The multiplication:

$$27 \times 59 = 1593$$

has the same last digit as:

$$7 \times 9 = 63$$

Experiment

Using the multiplication 58×17, try to see why the product of the last digits ends with the same digit as the product of 58 and 17. Write:

$$17 = 10 + 7$$

and use the distributive property as well as the fact that the product of a number by 10 ends in zero. Here is a bit of help:

$$
\begin{aligned}
58 \times (10 + 7) &= 580 + 58 \times 7 \\
58 \times 7 &= (50 + 8) \times 7 \\
7 \times (50 + 8) &= 350 + 7 \times 8
\end{aligned}
$$

The repeated multiplication of a digit by itself results in a repeating pattern of last digits.

2 = 2	3 = 3
2 x 2 = 4	3 x 3 = 9
2 x 2 x 2 = 8	3 x 3 x 3 = 27
2 x 2 x 2 x 2 = 16	3 x 3 x 3 x 3 = 81
2 x 2 x 2 x 2 x 2 = 32	3 x 3 x 3 x 3 x 3 = 243
2 x 2 x 2 x 2 x 2 x 2 = 64	3 x 3 x 3 x 3 x 3 x 3 = 729
2 x 2 x 2 x 2 x 2 x 2 x 2 = 128	3 x 3 x 3 x 3 x 3 x 3 x 3 = 2187

The last digits of the products form the sequence:
2, 4, 8, 6, 2, 4, 8, 6, ...

The last digits of the products form the sequence:
3, 9, 7, 1, 3, 9, 7, 1, ...

Experiment

- Multiply 4 by itself repeatedly. Show that the last digits of the successive products form the sequence: 4, 6, 4, 6, ...

- Multiply 5 by itself repeatedly. Show that the last digits of the successive products form the sequence: 5, 5, 5, 5, ...

- Multiply 6 by itself repeatedly. Show that the last digits of the successive products form the sequence: 6, 6, 6, 6, ...

- Multiply 7 by itself repeatedly. Show that the last digits of the successive products form the sequence: 7, 9, 3, 1, 7, 9, 3, ...

- Multiply 8 by itself repeatedly. Show that the last digits of the successive products form the sequence: 8, 4, 2, 6, 8, 4, 2, ...

- Multiply 9 by itself repeatedly. Show that the last digits of the successive products form the sequence: 9, 1, 9, 1, ...

Example: What is the last digit of the multiplication:

$$\underbrace{2 \times 2 \times 2 \times \cdots \times 2}_{222 \text{ times}}$$

When 2 is repeatedly multiplied by itself, the last digits form a sequence with the following terms:

$$2, \ 4, \ 8, \ 6, \ 2, \ 4, \ \ldots$$

Therefore, we ~~must~~ *must* find out how many groups of four terms there are:

$$222 = 55 \times 4 + 2$$

There are 55 groups of 4 terms. Each group has the following pattern of last digits: 2, 4, 8, 6. When 2 is multiplied by itself 220 times, the product ends in 6. When it is multiplied by itself 222 times, the product ends in 4.

$\checkmark$ PRACTICE TWO

Do not use a calculator for any of the problems!

Exercise 1

Find the last digit of the following products:

(A) $\underbrace{3 \times 3 \times 3 \times \cdots \times 3}_{299 \text{ factors}}$

(B) $\underbrace{4 \times 4 \times 4 \times \cdots \times 4}_{399 \text{ factors}}$

(C) $\underbrace{7 \times 7 \times 7 \times \cdots \times 7}_{699 \text{ factors}}$

(D) $\underbrace{9 \times 9 \times 9 \times \cdots \times 9}_{899 \text{ factors}}$

Exercise 2

Find the last digit of the following product:

$$\underbrace{4 \times 7 \times 4 \times 7 \times \cdots \times 7}_{200 \text{ factors}}$$

Exercise 3

If the last digit of the product

$$\underbrace{8 \times x \times 8 \times x \times \cdots \times x}_{100 \text{ factors}}$$

is 4, which digit is x?

Exercise 4

Make a list of the last digits when each digit from 0 to 9 is multiplied by itself.

Exercise 5

Find the last digit of the following product:

$$\underbrace{2 \times 3 \times 2 \times 3 \times \cdots \times 3}_{98 \text{ factors}}$$

Exercise 6

Find the last digit of the following product:

$$\underbrace{777 \times 777 \times \cdots \times 777}_{778 \text{ factors}}$$

Exercise 7

If the product

$$\underbrace{d \times d \times d \times d \times \cdots \times d}_{411 \text{ factors}}$$

ends in 7, which digit is d?

Exercise 8

If the product

$$\underbrace{7 \times 7 \times 7 \times \cdots \times 7}_{k \text{ factors}}$$

ends in 3, which of the following values cannot be k?

(A) 179

(B) 389

(C) 239

(D) 495

Exercise 9

Find the last digit of the following product:

$$2 \times 3 \times 4 \times 5 \times \cdots \times 20$$

Exercise 10

How many identical digits end the product:

$$1 \times 2 \times 3 \times 4 \times 5 \times \cdots \times 30$$

Exercise 11

In the result of the following product, what is the lowest place value with a non-zero digit? Place values are counted from right to left.

$$2 \times 4 \times 6 \times 8 \times 10 \times \cdots \times 50$$

(A) the 3rd place

(B) the 7th place

(C) the 8th place

(D) the 9th place

20

Exercise 12

Find the last three digits of the result of the following operation:

$$2 \times 4 \times 6 \times 8 \times 10 \times \cdots \times 50 - 69$$

Exercise 13

Find the last digit of the result of the following operation:

$$\underbrace{5 + 2 + 5 + 2 + 5 + 2 + \cdots + 5 + 2 + 5}_{201 \text{ terms}} =$$

Exercise 14

Find the last digit of the digit product of the following number:

$$\underbrace{24567892456789 \cdots 789}_{777 \text{ digits}}$$

THE PIGEONHOLE PRINCIPLE

If we place n objects in n − 1 boxes, there is at least one box with more than one object in it.

Experiment

Set aside 5 boxes and 6 toys. Close your eyes and place the toys in boxes as randomly as you can. Open your eyes and answer the questions:

- How many boxes are empty?
- How many boxes contain more than one toy?
- Can you move some of the toys from box to box so that there is no box with more than one toy in it?

Experiment

If you have 6 boxes and none of them contains more than one toy, what is the largest number of toys the boxes could contain in total?

Experiment

If you have 8 boxes and you place toys in them at random, at most how many toys should you place in order to have at least one box with 2 or more toys in it?

PRACTICE THREE

Do not use a calculator for any of the problems!

Exercise 1

Arbax, the Dalmatian, has 18 bones in 7 boxes. Which of the following must be true? Check all that apply.

(A) There are at least 4 boxes with 3 bones each.

(B) There is at least one box with 3 bones in it.

(C) There is at least one box with at least 3 bones in it.

(D) As many as 6 boxes could be empty.

Exercise 2

Lynda, the terrier, has a cache with 20 bones, some large and some small. Lynda cannot see how big a bone is before she pulls it out of the cache. Arbax asked her how many small bones she had and Lynda replied: "If I want to be sure I pull out a small bone, I have to take out at least 14 bones." Arbax was then able to figure out the answer to his question. Can you do the same?

Exercise 3

Dina, Lila, and Amira decided to help Alfonso sort out some vegetables. Alfonso had a mix of 20 peppers, 12 beets, and 18 zucchini in a large box. The girls counted the veggies.

1. Dina said: "If I removed vegetables one by one while blindfolded, I would have to remove at most .... vegetables before I could be sure that I had taken out at least one pepper."

2. Lila said: "If I removed vegetables one by one while blindfolded, I would have to remove at most .... vegetables before I could be sure that I had taken out at least one beet."

3. Amira said: "If I removed vegetables one by one while blindfolded, I would have to remove at most .... vegetables before I could be sure that I had taken out at least one pepper or one beet."

Fill in the missing numbers.

Exercise 4

While on the phone with a client, Stephan is reading his email and transferring tennis balls from a bag into a travel pouch without looking. In the bag, there are 50 Wilson balls, 30 Slazenger balls, and 25 Dunlop balls. How many balls must he transfer from the bag to the pouch to be sure he has at least 5 Wilson balls and 5 Slazenger balls?

Exercise 5

Amanda, the dance instructor, was teaching her students a new dance routine. Dina and Lila were among the students. There were 5 girls in an outer circle and 5 boys in a smaller inner circle. The girls took their positions but realized that none of them was aligned with her partner. How many of the following statements must be true:

(A) Dina said: "If we rotate the inner circle clockwise to the next dancer, I will meet my partner in less than 6 moves."

(B) Lila said: "If we do what you say, there will be a position, in less than 6 moves, in which at least two of us will be aligned with our partners."

(C) Amanda said: "It is not possible to have more than 1 dancer aligned with her partner in this way."

Exercise 6

Amira has a box of 40 crayons. The crayons come in five colors and there are at least 4 crayons of each color in the box. Amira plays a game with Dina. Blindfolded, each of them has to remove the crayons the other requests from the box. They must remove as few crayons as possible.

1. Amira asks: "Remove at least 3 crayons of the same color." How many crayons must Dina remove in total?

2. Dina asks: "Remove at least 3 red and 4 yellow crayons." How many crayons must Amira remove in total?

3. Amira asks: "Remove at least 4 crayons of a color and 4 crayons of another color." How many crayons must Dina remove in total?

Exercise 7

Lila participated in the Geography Bee. When she entered the exam room, she noticed that a row of 15 chairs had been reserved for the competitors. She also noticed that, wherever she chose to sit, she would have at least one person sitting beside her. What is the smallest number of competitiors that may have been seated at the time?

Exercise 8

Arbax and Lynda wanted to merge their bone collections. Lynda had 10 bones in 4 boxes and Arbax had 11 bones in 5 boxes.

1. At least how many boxes with at least 3 bones in each did Arbax have?

2. At least how many boxes with at least 3 bones in each did Lynda have?

3. At least how many boxes with at least 3 bones in each were in the merged collection?

Exercise 9

At the Old Planetarium, a robot makes red, blue, and green cards and, at random, stamps them with pictures of Jupiter, Saturn, and Neptune. It costs 10 cents to get a card printed. How much must Dina spend to be sure she obtains two cards with the same color and picture?

ARRANGEMENTS

Experiment

Place 5 *different* objects in 5 spots on a line. In how many ways can this be done? It is essential for the objects to be different. If the objects were identical, then there would be only one way to place them - just imagine 5 paperclips placed on a line!

There are $5 \times 4 \times 3 \times 2 \times 1 = \mathbf{120}$ **ways**:

There are 5 choices for the placement of the 1st object.

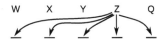

There are 4 choices for the placement of the 2nd object.

There are 3 choices for the placement of the 3rd object.

There are 2 choices for the placement of the 4th object.

There is 1 choice for the placement of the 5th object.

The product of the first 5 consecutive positive integers starting is called *five factorial*. The factorial notation (!) is simply a shorthand for such a product:

$$5! = 1 \times 2 \times 3 \times 4 \times 5$$

The product of the first N consecutive positive integers is called N *factorial*.

$$N! = 1 \times 2 \times 3 \times 4 \times \cdots \times N$$

The number of ways we can arrange N different objects on a line is $N!$

Even if we have zero objects, there is still one way of arranging them. Therefore:

$$0! = 1$$

Example 1:

Calculate the following values:

1. $2! =$

2. $3! =$

3. $4! =$

4. $5! = 120$

5. $6! =$

Example 2:

Place 5 different objects in 7 spots on a line. In how many ways can this be done?

Example 3:

Place 4 different objects in 3 spots on a line. In how many ways can this be done?

Answer 1:

1. $2! = 2$

2. $3! = 6$

3. $4! = 24$

4. $5! = 120$

5. $6! = 720$

Answer 2:

There are $7 \times 6 \times 5 \times 4 \times 3 = 42 \times 60 = 2520$ ways to do this.

Answer 3:

There are 4 ways of leaving one object out. There are 3! ways of placing the 3 chosen objects in the available spots. The total number of ways is $4 \times 6 = 24$.

Practice Four

Do not use a calculator for any of the problems!

Exercise 1

In how many ways can Lila, Dina, and Amira line up for movie tickets?

Exercise 2

Arbax and Lynda organized a dog race. They asked some cats to come over and encourage the racers to run after them. There are five dogs racing: Dee, Kay, Gee, Ess, and Tee. There are five cats they must run after: One, Two, Three, Four, and Five. Each dog pulls the name of the cat he will run after out of a hat. How many different choices are there?

Exercise 3

How many 3-digit numbers can be written using only different prime digits?

Exercise 4

Lila has 5 crayons of different colors. She wants to give Dina 2 crayons in exchange for a magnet. How many different pairs of crayons can Lila choose from?

Exercise 5

Amira has a string of 6 identical beads. In how many ways can she cut it into two smaller strings?

Exercise 6

How many 3-digit numbers have different digits?

Exercise 7

How many 3-digit numbers have exactly two identical digits?

Exercise 8

How many 3-digit numbers have at least two identical digits?

Exercise 9

How many 3-digit numbers have a first digit that is equal to the sum of the last two digits?

Exercise 10

How many 3-digit numbers have a last digit that is equal to the product of the first two digits?

Exercise 11

A robot paints cubes either red, green, or blue. Another robot organizes the painted cubes in groups of three. A third robot packages each group and labels it with its price. If red cubes cost 30 cents, blue cubes cost 40 cents, and green cubes cost 50 cents, how many different kinds of labels must the third robot be supplied with?

Exercise 12

A robot paints identical wooden sticks either green or red. Another robot picks out the painted sticks at random and uses them to make square frames. How many different types of frames can be manufactured in this way?

Exercise 13

For how many two digit numbers is the digit sum equal to the digit product?

Exercise 14

How many + symbols are there in the expression:

$$3 + 3 + 3 + \cdots + 3 = 21300$$

Exercise 15

Amira has drawn a flower with 11 petals. She wanted to paint some of the petals yellow. Are there more ways to paint 6 petals or more ways to paint 5 petals yellow?

Exercise 16

Alfonso is placing boxes of fresh fruit in his store's window. He has 5 boxes of mango and 5 boxes of apples and he wants to arrange them so that no two mango boxes are next to each other. How many different arrangements are there?

Exercise 17

Among her toys, Amira has 5 identical sloths and 5 identical manatees. She wants to arrange them in a circle so that no two sloths are next to each other. How many different arrangements are there?

Exercise 18

Four dogs have won prizes in the race organized by Arbax and Lynda: Als, Dux, Los, and Mew. Two dogs tied for the third place and the other two dogs won first and second place. How many possibilities are there for the placements of the four dogs?

Exercise 19

Four dogs won the race organized by Arbax and Lynda: Als, Dux, Los, and Mew. Two of the dogs have tied for one of the three top places. How many possibilities are there for the placements of the four dogs?

Exercise 20

Arbax has three bones: one small, one medium, and one large. He also has three caches: one by the pine tree, one by the shed, and one by the chickencoop. In how many different ways can he store the bones?

DOMINOES AND SQUARE TABLES

Dominoes are rectangular pieces made of two square halves. On each half, there are from 0 to 6 dots called *pips*. The rule for placing dominoes is to place ends with the same number of pips beside one another, as follows:

Many problems use dominoes to demonstrate various symmetry and counting concepts.

Square tables are grids with the same number of columns as rows.

Square tables are often populated with number sequences that bring a 2-dimensional twist to the problems.

Example What is the number at the center of the square table in the figure?

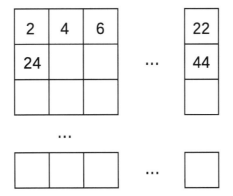

Based on the numbers shown in the first few cells, the table has $\dfrac{22-2}{2}+1=11$ columns. Since it is square, it also has 11 rows.

Because the dimensions of the table are odd, there is a center cell at the intersection of column 6 and row 6.

The number at row 1, column 6 is 12. From row to row, the numbers in the same column differ by 22. Therefore, the number in the center cell is $12+22\times5=122$.

Magic squares are a class of square tables in which the sum of the numbers on each row, each column, and each diagonal is the same. The number of rows (or columns) of a magic square is called *the order* of the magic square.

PRACTICE FIVE

Do not use a calculator for any of the problems!

Exercise 1

Answer the following questions about the square table:

1. How many cells does the table have?
2. Does the table have a cell at its center?
3. What is the number in the bottom right cell?

Exercise 2

Which number is in the center cell of the square table?

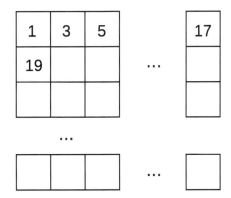

Exercise 3

Is it true that the sum of any 3 numbers we choose from the table below so that no two numbers are on the same column or row, is always the same?

n	n+a	n+2a
n+3a	n+4a	n+5a
n+6a	n+7a	n+8a

Exercise 4

A full set of dominoes has one domino for each possible unique pair of numbers of pips. How many dominoes are there in a full set?

Exercise 5

What is the sum of the numbers of pips on the shaded faces of the dominoes? (Dominoes from a single full set have been used.)

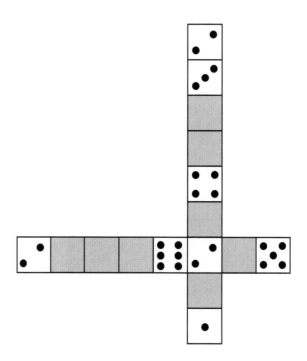

Exercise 6

Some numbers are missing from the following magic square. What is their sum?

4		12
		2
8		

MISCELLANEOUS PRACTICE

Do not use a calculator for any of the problems!

Exercise 1

How many 2-digit prime numbers have one digit that is twice the other?

Exercise 2

Max, the baker, can make six loaves from a 3 lbs bag of flour with enough flour left over to make another third of a loaf. How many bags of flour does he need in order to make 38 loaves?

Exercise 3

Amanda earns more money than she needs. Each month, she saves a third of her earnings and spends the rest. She worked for three years, during which her earnings and her expenses remained unchanged. If she stops working and keeps spending at the same rate, for how many months will her savings last?

Exercise 4

A robot produces white square tiles. Another robot produces black square tiles. A third robot picks 4 tiles at random and glues them to make a large square. How many different types of large squares is it possible to make?

Exercise 5

Ali and Baba have found a treasure chest filled with gold bars. They cannot use the gold bars to purchase what they need, since one bar is too valuable. They decided to cut the bars into coins. From each bar they can make 4 coins with some gold left over. The leftover gold from 6 bars can be used to make one whole bar. How many bars did they find if they were able to make 172 coins?

Exercise 6

From the pairs of sets defined below, select the pair that represents disjoint sets:

(A) the set of people on Dina's street who live at even numbered houses and the set of people who own dogs;

(B) the set of days with rain and the set of days with sun in Lila's town during August 2012;

(C) the set of mothers with only two children and the set of mothers with only three children on Amira's street;

(D) the set of houses with solar panels and the set of houses with fruit trees on Dina's street.

Exercise 7

Amanda is choreographing a new dance. In it, twelve dancers stand in a row, equally spaced, waving scarves. Amanda then dances from one end of the row to the other end, passing in front of each dancer. It takes Amanda 4 seconds to dance from the first dancer to the fifth dancer. How many seconds does it take her to dance from the one end of the row to the other end of the row?

Exercise 8

What is the last digit of the product:

$$3 \times 33 \times 333 \times 3333 \times \cdots \times \underbrace{33\cdots3}_{111 \text{ digits}}$$

Exercise 9

Of the following pairs of sets, which ones represent a set that is included in the other? Check all that apply.

(A) the set of dogs who won the race and the set of dogs who trained for the race;

(B) the set of prime numbers and the set of fractions;

(C) the set of vegetables and the set of boxes in Alfonso's store;

(D) the set of cars on the road and the set of passengers in those cars.

Exercise 10

How many different ways of painting a square black are there in a 4×4 grid of white squares?

Exercise 11

Enumerate the elements of each set:

1. the set of prime numbers smaller than 20;

2. the set of planets in the solar system;

3. the set of (possibly meaningless) words that can be formed with the letters M, E, and W;

4. the set of playing cards that are multiples of 5.

Exercise 12

How many elements does each set have?

1. the set of factors of 441;

2. the set of students from the 4th in line to the 24th in line;

3. the set of digits used to write integer numbers from 1 to 99;

4. the set of 5 digit numbers with a digit sum of 4 and a non-zero digit product.

Exercise 13

How many rectangles with integer sides have a perimeter of 14?

Exercise 14

What are the last 3 digits of the product:

$$20! = 1 \times 2 \times 3 \times \cdots \times 20$$

Exercise 15

Amira has lots of magnetic balls and rods. She made the following pattern:

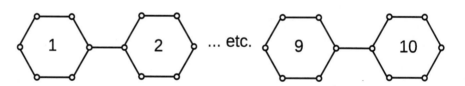

How many balls and how many rods did she use?

Exercise 16

In which of the following rectangular grids is it possible to color a third of the small squares black?

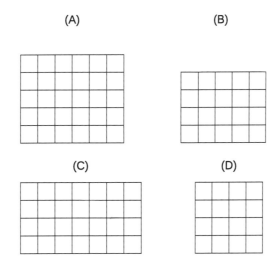

Exercise 17

How many digits of 0 are there in the result of the operation:

$$\underbrace{444\cdots44}_{444 \text{ digits}}\times7$$

Exercise 18

Lila's math club walked to the school's parking lot for an experiment. Lila noticed that there were 28 blue cars in the parking lot. Of these, 15 had tinted windows and 19 had roofracks. What could have been the largest possible number of blue cars with tinted windows but without roofracks?

Exercise 19

How many 2-digit numbers are divisible by their tens digit?

Exercise 20

Amira had some red and some blue blocks in a box. If Lila takes some blocks out of the box blindfolded, she has to take out at least 11 blocks to be sure she has at least 2 blue blocks and she has to take out at least 8 blocks to be sure she has at least 3 red blocks. How many blocks did Amira have in the box?

Exercise 21

Which of the following arrangements are possible using a full set of dominoes?

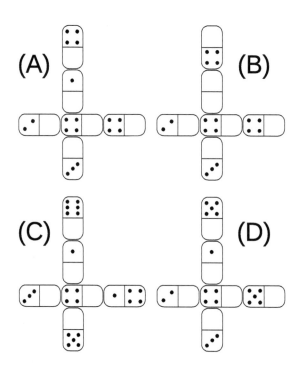

Exercise 22

Dina and Lila have 2 red chickens, one white chicken, and one black chicken. When the sun goes down, the chickens retire to their nests where they sit in a line. In how many different ways can they sit if the red chickens always sit side by side?

Exercise 23

Alfonso has to place some egg cartons in the store's window for advertisement. He has two boxes labeled "cage free," one box labeled "free range," and one box labeled "happy hens." In how many ways can he make a row of boxes out of them if the two "cage free" boxes have to be placed side by side?

Exercise 24

The three machines in the figure can be combined to produce an output when a number is input. How many different results can we obtain from a single input if we are allowed to use each machine once but in any order?

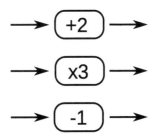

Exercise 25

How many rectangles are there in the figure?

Exercise 26

How many rectangles are there in the figure?

1	2		20

Exercise 27

How many rectangles are there in the figure?

Exercise 28

In how many different ways can we place the black rectangle inside the white rectangle so that it covers two squares exactly?

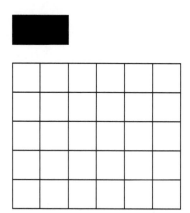

Assume the large grid cannot be rotated or mirrored.

SOLUTIONS TO PRACTICE ONE

Exercise 1

Draw a Venn diagram that represents the following sets:

- the set of tennis balls at Stephan's club
- the set of Wilson tennis balls at Stephan's club
- the set of pink tennis balls at Stephan's club

Solution 1

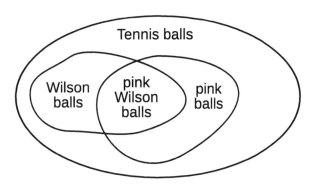

Exercise 2   (i) Draw a Venn diagram that represents the following sets:

- Set W: the set of vegetables Max displayed in the store this morning
- Set X: the set of potatoes Max displayed in the store this morning
- Set Y: the set of purple potatoes Max displayed in the store this morning
- Set Z: the set of yellow flesh potatoes Max displayed in the store this morning

(ii) Which of the sets in your diagram are disjoint?

(A) set Z (yellow flesh potatoes) and set X (potatoes)

(B) set Y (purple potatoes) and set W (vegetables)

(C) set Y (purple potatoes) and set Z (yellow flesh potatoes)

(D) set W (vegetables) and set X (potatoes)

(iii) In your diagram, which sets are subsets of the set of vegetables?

Solution 2

1. The Venn diagram is:

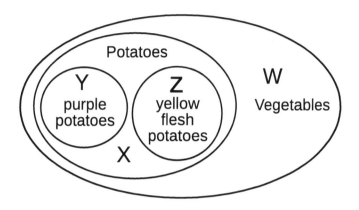

2. The correct answer is (C).

3. Sets X, Y, and Z are subsets of the set W.

Exercise 3

Dina went to a wilderness survival class. The girls and boys in the class went to gather mushrooms. Some of them were lucky and found mushrooms and some of them did not. Among the students who found mushrooms, some were girls and some were boys. Make a Venn diagram that illustrates the various sets and subsets in this story.

Fill in the blanks in the paragraph.

Solution 3

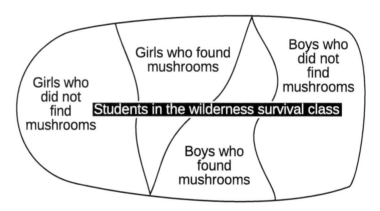

The set of boys is a *subset* of the set of students in the survival class. The set of girls is a *subset* of the set of students in the survival class. The set of boys and the set of girls are *disjoint*. The set of girls and boys who found mushrooms is a *subset* of the set of students in the survival class.

Exercise 4

Which of the following describes two subsets that must be disjoint? Check all that apply.

(A) At 1 PM, there were 18 cars in the parking lot at Alfonso's grocery. 4 cars were blue and 14 cars had tinted windows.

(B) At 1 PM, there were 18 cars in the parking lot at Alfonso's grocery. 5 cars had license plates that started with the letter Z and 3 cars had license plates that started with the letter Y.

(C) At 1 PM, there were 18 cars in the parking lot at Alfonso's grocery. 3 cars belonged to staff and 14 cars belonged to customers.

(D) At 1 PM, there were 18 cars in the parking lot at Alfonso's grocery. 8 cars were hatchbacks and 10 cars had roof racks.

Solution 4

Descriptions (B) and (C) correspond to disjoint subsets of the set of cars. Descriptions (A) and (D) are not conclusive.

(A) Some of the blue cars may have tinted windows. It is not clear that the subsets are disjoint. While it is true that the number of elements in the two subsets totals 18, it is possible that there are some cars that are not blue and do not have tinted windows.

(B) If a license plate starts with Z, it cannot also start with X. The two subsets must be disjoint. While it is true that the number of elements in the two subsets does not total 18, it is possible that there are cars with license plates that start with other letters.

(C) If a car belongs to staff, it cannot belong to a customer. The subsets must be disjoint. While it is true that the number of elements of the two subsets does not total 18, it is possible that a taxi or a police car may be parked in the lot.

(D) Even though the number of elements of the two subsets totals 18, the subsets need not be disjoint. Some hatchback cars may also have roof racks, while other cars in the lot may have neither a roof rack nor a hatchback.

Exercise 5

In a deck of playing cards, which of the following are disjoint subsets? Check all that apply.

(A) the picture cards and the diamonds

(B) the diamonds and the 4s

(C) the picture cards and the 4s

(D) the clubs and the diamonds

Solution 5

(A) The picture cards and the diamonds are not disjoint. Some picture cards are diamonds: for example, the queen of diamonds.

(B) The diamonds and the 4s are not disjoint. There is a 4 of diamonds.

(C) The picture cards and the 4s are disjoint. There are no picture cards that are also 4s.

(D) The clubs and the diamonds are disjoint.

Exercise 6

Amira's class went on a field trip to the Museum of Art. Each of the 24 students in the class either used an audioguide or a notepad.If 15 used an audioguide and 18 took notes in their notepads, how many students used an audioguide and also took notes?

Solution 6

Since $15 + 18 = 33$ is larger than the total number of students by 9, 9 students used both an audioguide and a notepad.

Exercise 7

Max, the baker, made 48 pieces of pastry this morning. Of these, 30 contain chocolate and 25 contain almonds. How many pieces of pastry contain both chocolate and almonds?

Solution 7

Since $30 + 25 = 55$ exceeds 48 by 7, there must be 7 pieces of pastry that contain both almonds and chololate.

Exercise 8

Cornelia has 30 ducks on her farm. Every second day, 16 of them each lay an egg. The rest each lay an egg every third day. What is the largest number of eggs Cornelia may collect in one week?

Solution 8

The ducks that lay an egg every second day may lay as many as 4 eggs each in a single week. The ducks that lay an egg every third day may lay as many as 3 eggs each in a single week. The largest number of eggs Cornelia may harvest in one week is:

$$16 \times 4 + 14 \times 3 = 64 + 42 = 106$$

Exercise 9

Stephan, the tennis coach, received 44 applications for a mock doubles tournament. In a doubles game, there are either 4 female players or 4 male players. Stephan must admit or reject each application. If each participant plays only once in the preliminary round, what is the largest number of applicants Stephan may have to reject?

Solution 9

We are not told how many of the applicants are female and how many are male. In the worst case, neither the number of males nor the number of females is a multiple of 4. Stephan will have to reject some applicants because he cannot assign them to teams. The number of women can be:

Case 1: a multiple of 4. In this case, since the total number of applicants is a multiple of 4, the number of men is also a multiple of 4. Stephan will not reject anyone.

Case 2: 1 larger than a multiple of 4. In this case, the number of men will be 3 larger than a multiple of 4. Stephan will have to reject 4 people: one woman and three men.

Case 3: 2 larger than a multiple of 4. In this case, the number of men will be 2 larger than a multiple of 4. Stephan will have to reject 4 people: two women and two men.

Case 4: 3 larger than a multiple of 4. In this case, the number of men will be 1 larger than a multiple of 4. Stephan will have to reject 4 people: three women and one man.

Exercise 10

Patrick, the leader of the math club at Amira's school, asked the club members for their sizes, so he could order T-shirts with the club logo for them. The club members gave him their sizes in the form of a riddle: "All of us are size S, except for six of us. All of us are size M, except for seven of us. All of us are size L, except for nine of us." How many members are in the club have? How did Patrick figure out how many T-shirts of each size he should order?

Solution 10

Six students wear M or L, seven students wear L or S, and nine students wear S or M:

$$
\begin{aligned}
L + M &= 6 \\
S + L &= 7 \\
M + S &= 9
\end{aligned}
$$

Add the three equations together:

$$
\begin{aligned}
S + S + M + M + L + L &= 22 \\
S + M + L &= 11
\end{aligned}
$$

and:

$$
\begin{aligned}
S = S + M + L - (M + L) = 11 - 6 &= 5 \\
M = S + M + L - (S + L) = 11 - 7 &= 4 \\
L = S + M + L - (M + S) = 11 - 9 &= 2
\end{aligned}
$$

There are 11 students in the math club. The leader was able to figure out the number of T-shirts of each size.

Exercise 11

In the following Venn diagram:

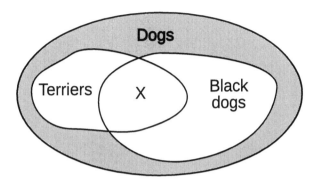

1. How would you describe the set X?

2. Make a list of the sets that are subsets of other sets. How many items are there?

3. How would you describe the shaded part of the set "Dogs?"

Solution 11

1. X is the set of black terriers.

2. There are 5 items in the list:

   - Terriers are a subset of Dogs.
   - Black Dogs are a subset of Dogs.
   - Black Terriers is a subset of Terriers.
   - Black Terriers is a subset of Black Dogs.
   - Black Terriers is a subset of Dogs.

3. Dogs that are neither terriers nor black.

SOLUTIONS TO PRACTICE TWO

Exercise 1

Find the last digit of the following products:

(A) $\underbrace{3 \times 3 \times 3 \times \cdots \times 3}_{299 \text{ times}}$

(B) $\underbrace{4 \times 4 \times 4 \times \cdots \times 4}_{399 \text{ times}}$

(C) $\underbrace{7 \times 7 \times 7 \times \cdots \times 7}_{699 \text{ times}}$

(D) $\underbrace{9 \times 9 \times 9 \times \cdots \times 9}_{899 \text{ times}}$

Solution 1

(A) $299 = 74 \times 4 + 3$ The product ends in 7.

(B) $399 = 199 \times 2 + 1$ The product ends in 4.

(C) $699 = 174 \times 4 + 3$ The product ends in 3.

(D) $899 = 449 \times 2 + 1$ The product ends in 9.

Exercise 2

Find the last digit of the following product:

$$\underbrace{4 \times 7 \times 4 \times 7 \times \cdots \times 7}_{200 \text{ factors}}$$

Solution 2

Strategy 1:

The product has 100 factors of 4 and 100 factors of 7. 100 is a multiple of both 2 and 4. Therefore, the product of 100 7s ends in 1 and the product of 100 4s ends in 6. Their product ends in 6.

Strategy 2:

There are 49 pairs of products 2×3. Each such product ends in 6. The product of any number of numbers that end in 6 ends in 6.

Exercise 3

If the last digit of the product:

$$\underbrace{8 \times x \times 8 \times x \times \cdots \times x}_{100 \text{ factors}}$$

is 4, which digit is x?

Solution 3

The product has 50 factors of 8 and 50 factors of x. The product of the 50 factors of 8 ends in 4. The product of 50 factors of x must end in 1. The only digits that give products ending in 1 when multiplied repeatedly are 1, 3, 7 and 9. Repeated multiplication of 7 produces a last digit of 1 every 4 terms. Since 50 is not a multiple of 4, x cannot be 7. Repeated multiplication of 3 produces a last digit of 1 every 4 terms as well - therefore, x cannot be 3. Repeated multiplication of 9 produces a last digit of 1 every second term. Since 50 is even, x can be 9. Obviously, x can also be 1.

Answer: x can be either 1 or 9.

Exercise 4

Make a list of the last digits when each digit from 0 to 9 is multiplied by itself.

Solution 4

product	0×0	1×1	2×2	3×3	4×4	5×5	6×6
		9×9	8×8	7×7			
ends in	0	1	4	9	6	5	6

The digits 2, 3, 7, and 8 are not in the list.

Exercise 5

Find the last digit of the following product:

$$\underbrace{2 \times 3 \times 2 \times 3 \times \cdots \times 3}_{98 \text{ factors}}$$

Solution 5

Strategy 1: There are 49 pairs of products 2×3. Each product 2×3 ends in 6. Repeated multiplications of 6 always end in 6.

The product ends in 6.

Strategy 2: There are 49 multiplications by 2 and 49 multiplications by 3. The repeated multiplications of 2 end with the digits:

$$2, \ 4, \ 8, \ 6, \ 2, \ 4, \ \ldots$$

The sequence repeats every 4 terms. Since $49 = 12 \times 4 + 1$, the product of 49 factors of 2 ends in 2.

The repeated multiplications of 3 end with the digits:

$$3, \ 9, \ 7, \ 1, \ 3, \ 9, \ \ldots$$

The sequence repeats every 4 terms. Since $49 = 12 \times 4 + 1$, the product of 49 factors of 3 ends in 3.

The product ends in 6.

Exercise 6

Find the last digit of the following product:

$$\underbrace{777 \times 777 \times \cdots \times 777}_{778 \text{ factors}}$$

Solution 6

The last digit of the product is the last digit of the product of the last digits. Repeated multiplications of 7 end with the following digits:

$$7, \ 9, \ 3, \ 1, \ 7, \ 9, \ \ldots$$

The sequence repeats every 4 terms. Since $778 = 194 \times 4 + 2$, the product ends in 9.

Exercise 7

If the product:

$$\underbrace{d \times d \times d \times d \times \cdots \times d}_{411 \text{ factors}}$$

ends in 7, which digit is d?

Solution 7

Only repeated multiplications of $d = 3$ and $d = 7$ can end in 7. The last digit sequences are, respectively:

$$3, \ 9, \ 7, \ 1, \ 3, \ 9, \ \ldots$$

$$7, \ 9, \ 3, \ 1, \ 7, \ 9, \ \ldots$$

Since $411 = 102 \times 4 + 3$, only the upper sequence satisfies as it has 7 occuring every third term after a multiple of 4. d is equal to 3.

Exercise 8

If the product

$$\underbrace{7 \times 7 \times 7 \times \cdots \times 7}_{k \text{ factors}}$$

ends in 3, which of the following values cannot be k?

(A) 179

(B) 389

(C) 239

(D) 495

Solution 8

Repeated multiplications of 7 end as follows:

$$7, \ 9, \ 3, \ 1, \ 7, \ 9, \ \ldots$$

The last digit can be 3 only if the number of factors is three more than a multiple of 4.

The answer choices can be written as:

(A) $179 = 44 \times 4 + 3$

(B) $389 = 97 \times 4 + 1$

(C) $239 = 59 \times 4 + 3$

(D) $495 = 123 \times 4 + 3$

The correct answer is (B).

Exercise 9

Find the last digit of the following product:

$$2 \times 3 \times 4 \times 5 \times \cdots \times 20$$

60

Solution 9

Whenever a factor of 2 and a factor of 5 can be paired, there is a multiplication by 10. This adds a zero at the end of the number.

Since there are many factors of 2 and of 5 among the numbers which are multiplied together, the last digit of the product is 0.

Exercise 10

How many identical digits end the product:

$$1 \times 2 \times 3 \times 4 \times 5 \times \cdots \times 30$$

Solution 10

Whenever a factor of 2 and a factor of 5 can be paired, there is a multiplication by 10. This adds a zero at the end of the number. Therefore, the problem asks how many zeros the number ends in.

To find out the number of zeros, we have to count the number of (2, 5) pairs that can be found among the factors. First, we notice that there are many more factors of 2 than of 5, since an even number occurs every second number while a multiple of 5 occurs only every five numbers. The number of (2, 5) pairs is, therefore, equal to the number of factors of 5.

There are factors of 5 every 5 consecutive numbers, but there is also an additional factor of 5 that occurs every 25 numbers:

$$\mathbf{5,\ 2 \times 5,\ 3 \times 5,\ 4 \times 5,\ 5 \times 5,\ 6 \times 5}$$

From 1 to 30 inclusive, there are 7 factors of 5, therefore the number ends with 7 zeroes.

Exercise 11

In the result of the following product, what is the lowest place value with a non-zero digit? Place values are counted from right to left.

$$2 \times 4 \times 6 \times 8 \times 10 \times \cdots \times 50$$

(A) the 3rd place

(B) the 7th place

(C) the 9th place

(B) the 8th place

Solution 11

The number of factors of 5 will tell us how many zeros the number ends with.

There is one factor of 5 in 10, 20, 30, 40, and 50. Additionally, 50 has another factor of 5 since it is a multiple of 25. There are 6 factors of 5 in total.

The smallest place with a non-zero digit is the 7th place. The correct answer is (B).

Exercise 12

Find the last three digits of the result of the following operation:

$$2 \times 4 \times 6 \times 8 \times 10 \times \cdots \times 50 - 69$$

Solution 12

As we have seen in the previous exercise, when computed, the product ends in 6 zeroes. After subtracting 69 from the product, the result will end in 931. Note that we do not need to compute the product!

Exercise 13

Find the last 2 digits of the result of the following operation:

$$\underbrace{5 + 2 + 5 + 2 + 5 + 2 + \cdots + 5 + 2 + 5}_{201 \text{ terms}} =$$

Solution 13

There are 100 $2 + 5$ pairs and an extra 5. 100 times any number ends in 00. If we add a 5 to this product, the last two digits are 05.

Exercise 14

Find the last digit of the digit product of the following number:

$$\underbrace{24567892456789 \cdots 789}_{777 \text{ digits}}$$

Solution 14

Since there is at least one digit of 2 and at least one digit of 5, the last digit of the digit product is zero. The additional information in the statement is not relevant.

Solutions to Practice Three

Exercise 1

Arbax, the Dalmatian, has 18 bones in 7 boxes. Which of the following must be true? Check all that apply.

(A) There are at least 4 boxes with 3 bones each.

(B) There is at least one box with 3 bones in it.

(C) There is at least one box with at least 3 bones in it.

(D) As many as 6 boxes could be empty.

Solution 1

Only (C) and (D) must be true.

(A) is not necessarily true, because it is possible to have the following distribution of bones: 4, 4, 2, 2, 2, 2, 2. Other distributions that invalidate (A) are possible.

(B) is not necessarily true, because we can easily find a counter-example. The example used to invalidate (A) also invalidates (B).

(C) is true.

(D) is true. Arbax could place all the bones in a single box.

Exercise 2

Lynda, the terrier, has a cache with 20 bones, some large and some small. Lynda cannot see how big a bone is before she pulls it out of the cache. Arbax asked her how many small bones she had and Lynda replied: "If I want to be sure I pull out a small bone, I have to take out at least 14 bones." Arbax was then able to figure out the answer to his question. Can you do the same?

Solution 2

There must be 13 large bones and 7 small bones in the cache. In the most unfavorable case, Lynda will take out all the large bones first and only remove a small bone on the 14th attempt, when there are only small bones left.

Exercise 3

Dina, Lila, and Amira decided to help Alfonso sort out some vegetables. Alfonso had a mix of 20 peppers, 12 beets, and 18 zucchini in a large box. The girls counted the veggies.

1. Dina said: "If I removed vegetables one by one while blindfolded, I would have to remove at most .... vegetables before I could be sure that I had taken out at least one pepper."

2. Lila said: "If I removed vegetables one by one while blindfolded, I would have to remove at most .... vegetables before I could be sure that I had taken out at least one beet."

3. Amira said: "If I removed vegetables one by one while blindfolded, I would have to remove at most .... vegetables before I could be sure that I had taken out at least one pepper or one beet."

Fill in the missing numbers.

Solution 3

1. Dina said: "If I removed vegetables one by one while blindfolded, I would have to remove at most 30 vegetables before I could be sure that I had taken out at least one pepper."

2. Lila said: "If I removed vegetables one by one while blindfolded, I would have to remove at most 28 vegetables before I could be sure that I had taken out at least one beet."

3. Amira said: "If I removed vegetables one by one while blindfolded, I would have to remove at most 18 vegetables before I could be sure that I had taken out at least one pepper or one beet."

Exercise 4

While on the phone with a client, Stephan is reading his email and transferring tennis balls from a bag into a travel pouch without looking. In the bag, there are 50 Wilson balls, 30 Slazenger balls, and 25 Dunlop balls. How many balls must he transfer from the bag to the pouch to be sure he has at least 5 Wilson balls and 5 Slazenger balls?

Solution 4

In the worst case, Stephan will transfer all the Wilson and Dunlop balls before transferring any Slazenger balls. He has to transfer $25 + 50 + 5 = 80$ balls.

Exercise 5

Amanda, the dance instructor, was teaching her students a new dance routine. Dina and Lila were among the students. There were 5 girls in an outer circle and 5 boys in a smaller inner circle. The girls took their positions but realized that none of them was aligned with her partner. How many of the following statements must be true:

(A) Dina said: "If we rotate the inner circle clockwise to the next dancer, I will meet my partner in less than 6 moves."

(B) Lila said: "If we do what you say, there will be a position, in less than 6 moves, in which at least two of us will be aligned with our partners."

(C) Amanda said: "It is not possible to have more than 1 dancer aligned with her partner in this way."

Solution 5

Two statements are true and one is false.

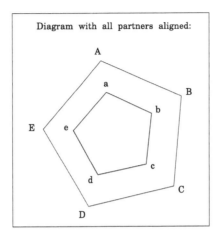

Dina is correct. Apart from the dancer she is aligned with currently, who is not her partner, there are 4 other dancers in the inner circle. After one of the next 4 moves, she will necessarily be aligned with her partner. This is true for any of the 5 dancers in one of the circles.

Lila is correct. There are only 4 moves possible before returning to the original position where there are no matches. Since there are 5 dancers and 4 possible moves, there must be a move after which at least 2 dancers are aligned with their partners.

Amanda's statement contradicts Lila's. It is thus false.

A	B	C	D	E	
b	d	e	a	c	sample starting position
d	e	a	c	b	
e	a	c	b	d	
a	c	b	d	e	sample line with 2 matches
c	b	d	e	a	

67

Exercise 6

Amira has a box of 40 crayons. The crayons come in five colors and there are at least 4 crayons of each color in the box. Amira plays a game with Dina. Blindfolded, each of them has to remove the crayons the other requests from the box. They must remove as few crayons as possible.

1. Amira asks: "Remove at least 3 crayons of the same color." How many crayons must Dina remove in total?

2. Dina asks: "Remove at least 3 red and 4 yellow crayons." How many crayons must Amira remove in total?

3. Amira asks: "Remove at least 4 crayons of a color and 4 crayons of another color." How many crayons must Dina remove in total?

Solution 6

1. Dina must remove 11 crayons. Since there are 5 different colors of crayons, it is possible that Dina might pick out 2 crayons of each color. She then has to pick out one more crayon to be sure that she has remove 3 crayons of the same color.

2. Amira must remove all 40 crayons. It is possible that there are only 4 red and 4 yellow crayons in the box. Amira could pull out as many as $40 - 8 = 32$ crayons that are neither red nor yellow. The next 4 crayons she picks out could all be red. Therefore, she must pull out all of the remaining 8 crayons to be sure she has removed 4 yellow crayons.

3. Dina has to remove 37 crayons. As many as $40 - 4 \times 4 = 24$ crayons could have the same color. Dina could pull these out first without having removed any crayons of a different color. From the remaining 16 crayons, Dina could remove as many as 12 crayons and still not have 4 crayons of a different color. Dina would have to remove one more crayon. $24 + 12 + 1 = 37$.

Exercise 7

Lila participated in the Geography Bee. When she entered the exam room, she noticed that a row of 15 chairs had been reserved for the competitors. She also noticed that, wherever she chose to sit, she would have at least one person sitting beside her. What is the smallest number of competitiors that may have been seated at the time?

Solution 7

5 competitors, seated as follows:

Exercise 8

Arbax and Lynda wanted to merge their bone collections. Lynda had 10 bones in 4 boxes and Arbax had 11 bones in 5 boxes.

1. At least how many boxes with at least 3 bones in each did Arbax have?

2. At least how many boxes with at least 3 bones in each did Lynda have?

3. At least how many boxes with at least 3 bones in each were in the merged collection?

Solution 8

1. Arbax had at least one box with at least 3 bones in it.

2. Lynda had at least one box with at least 3 bones in it.

3. The merged collection had at least one box with at least 3 bones in it.

Exercise 9

At the Old Planetarium, a robot makes red, blue, and green cards and, at random, stamps them with pictures of Jupiter, Saturn, and Neptune. It costs 10 cents to get a card printed. How much must Dina spend to be sure she obtains two cards with the same color and picture?

Solution 9

There are 9 (3 × 3) distinct cards. If Dina purchases 10 cards, there are sure to be two cards of the same kind in the set. Dina must spend one dollar in total.

SOLUTIONS TO PRACTICE FOUR

Exercise 1

In how many ways can Lila, Dina, and Amira line up for movie tickets?

Solution 1

In $3! = 6$ ways:

- Dina, Lila, Amira
- Dina, Amira, Lila
- Amira, Dina, Lila
- Amira, Lila, Dina
- Lila, Dina, Amira
- Lila, Amira, Dina

Exercise 2

Arbax and Lynda organized a dog race. They asked some cats to come over and encourage the racers to run after them. There are five dogs racing: Dee, Kay, Gee, Ess, and Tee. There are five cats they must run after: One, Two, Three, Four, and Five. Each dog pulls the name of the cat he will run after out of a hat. How many different choices are there?

Solution 2

$5! = 120$. The first dog to pull out a cat name has 5 choices, the second dog has 4 choices, etc.

Exercise 3

How many 3-digit numbers can be written using only different prime digits?

Solution 3

The prime digits are: 2, 3, 5, and 7. There are 4 ways of choosing which digit to leave out in order to be left with 3 digits that form the number. Since the 3 chosen digits are different, there are $3! = 6$ ways to put them in the units, tens, and hundreds places. For example, leave 5 out and form numbers with the digits 2, 3, and 7. Six different numbers can be formed:

$$237, \ 273, \ 327, \ 372, \ 723, \ 732$$

A total of 24 distinct 3-digit numbers can be formed.

Exercise 4

Lila has 5 crayons of different colors. She wants to give Dina 2 crayons in exchange for a magnet. How many different pairs of crayons can Lila choose from?

Solution 4

Let us assume the colors are R, G, B, Y, O. Lila can pair any crayon with 4 other crayons. When counting the number of pairs, we must make sure we do not count any pair twice. For example, when we form the pairs O-R, O-G, O-B, and O-Y and compare them with the pairs R-G, R-B, R-Y, and R-O, we see that the pair R-O has been counted twice. If we write down all 5×4 pairs, we notice that each pair is counted twice. Lila can choose from 10 different pairs of crayons.

Exercise 5

Amira has a string of 6 identical beads. In how many ways can she cut it into two smaller strings?

Solution 5

There are three possible ways:

$$
\begin{aligned}
6 &= 1 + 5 \\
6 &= 2 + 4 \\
6 &= 3 + 3
\end{aligned}
$$

Exercise 6

How many 3-digit numbers have different digits?

Solution 6

There are 10 digits in total.

- The leftmost digit cannot be zero, since the number would then be a 2-digit number. There are 9 choices for this digit.
- The middle digit can be zero, but it cannot be the same digit as the leftmost digit. There are 9 choices for this digit.
- The rightmost digit can be zero, but it cannot be the same as either of the other digits. There are 8 choices for this digit.

In total, there are $8 \times 9 \times 9$ 3-digit numbers with different digits.

$$
\begin{aligned}
8 \times 9 &= 72 \\
72 \times 9 &= 72 \times 10 - 72 \\
720 - 72 &= 648
\end{aligned}
$$

Exercise 7

How many 3-digit numbers have exactly two identical digits?

Solution 7

The identical digits can be in three different places:

Case 1: $\underline{Z} \quad \underline{W} \quad \underline{Z}$

Case 2: $\underline{Z} \quad \underline{Z} \quad \underline{W}$

Case 3: $\underline{W} \quad \underline{Z} \quad \underline{Z}$

The first (leftmost) digit can never be zero.

(a) **Case 1:** There are 9 possible choices for Z (digits from 1 to 9) and 9 possible choices for W (digits from 0 to 9 but not the same digit as Z). Total: 81 numbers.

(b) **Case 2:** As in case 1, there are 81 choices for this pattern.

(c) **Case 3:** There are 9 possible choices for W (digits from 1 to 9) and 9 possible choices for Z (digits from 0 to 9 but not the same digit as W). Total: 81 numbers.

There are $81 \times 3 = 243$ numbers with the required property.

Exercise 8

How many 3-digit numbers have at least two identical digits?

Solution 8

This problem is similar to the previous problem, but now we must include the numbers that have 3 identical digits. There are 9 numbers with 3 identical digits (any digit from 1 to 9). Add this to the result of the previous problem to get $243 + 9 = 252$. There are 252 numbers with the required property.

Exercise 9

How many 3-digit numbers have a first digit that is equal to the sum of the last two digits?

Solution 9

The sum of the last two digits cannot exceed 9:

Strategic Solution:

The first digit can have any value from 1 to 9. For each value, the remaining two digits must add up to it. The number of possibilities is equal to one more than the value of the first digit. Just imagine transfering 1 from one term to the other. Take, for example, a first digit of 5:

$$5 = 0 + 5$$
$$5 = 1 + 4$$
$$5 = 2 + 3$$
$$5 = 3 + 2$$
$$5 = 4 + 1$$
$$5 = 5 + 0$$

This must be true for any value of the first digit, so the number of numbers is:

$$2 + 3 + 4 + \cdots + 10 =$$

Let us compute this by using a triangular number:

$$2 + 3 + 4 + \cdots + 10 = 1 + 2 + 3 + \cdots + 10 - 1 = \frac{10 \times 11}{2} - 1 = 55 - 1 = 54$$

There are 54 numbers with the required property.

Brute Force Solution:

- First digit is equal to 1: the last two digits can be 01 and 10. There are 2 such numbers.

- First digit is equal to 2: the last two digits can be 02, 20, and 11. There are 3 such numbers.

- First digit is equal to 3: the last two digits can be 03, 30, 21, and 12. There are 4 such numbers.

- First digit is equal to 4: the last two digits can be 04, 40, 31, 13, and 22. There are 5 such numbers.

- First digit is equal to 5: the last two digits can be 05, 50, 41, 14, 23, and 32. There are 6 such numbers.

- First digit is equal to 6: the last two digits can be 06, 60, 51, 15, 24, 42, and 33. There are 7 such numbers.

- First digit is equal to 7: the last two digits can be 07, 70, 61, 16, 25, 52, 34 and 43. There are 8 such numbers.

- First digit is equal to 8: the last two digits can be 08, 80, 71, 17, 26, 62, 53, 35, and 44. There are 9 such numbers.

75

- First digit is equal to 9: the last two digits can be 09, 90, 81, 18, 27, 72, 45, and 54. There are 10 such numbers.

Compute the sum:

$$2 + 3 + 4 + \cdots + 10 =$$

by using a triangular number:

$$2+3+4+\cdots+10 = 1+2+3+\cdots+10-1 = \frac{10 \times 11}{2} - 1 = 55 - 1 = 54$$

There are 54 numbers with the required property.

Exercise 10

How many 3-digit numbers have a last digit that is equal to the product of the first two digits?

Solution 10

The last digit may be equal to zero if the middle digit is zero. In this case, the first digit can have any value between 1 and 9. There are 9 numbers that fulfill this requirement.

The last digit may also be:

$1 = 1 \times 1$ 1 number fulfills: 111;

$2 = 1 \times 2$ 2 numbers fulfill: 122 and 212;

$3 = 1 \times 3$ 2 numbers fulfill: 133 and 313;

$4 = 1 \times 4$ 2 numbers fulfill: 144 and 414;

$4 = 2 \times 2$ 1 number fulfills: 224;

$5 = 1 \times 5$ 2 numbers fulfill: 155 and 515;

$6 = 1 \times 6$ 2 numbers fulfill: 166 and 616;

$6 = 2 \times 3$ 2 numbers fulfill: 236 and 326;

$7 = 1 \times 7$ 2 numbers fulfill: 177 and 717;

$8 = 1 \times 8$ 2 numbers fulfill: 188 and 818;

$8 = 2 \times 4$ 2 numbers fulfill: 248 and 428;

$9 = 1 \times 9$ 2 numbers fulfill: 199 and 919;

$9 = 3 \times 3$ 1 number fulfills: 339;

There are $23 + 9 = 32$ numbers with the required property.

Exercise 11

A robot paints cubes either red, green, or blue. Another robot organizes the painted cubes in groups of three. A third robot packages each group and labels it with its price. If red cubes cost 30 cents, blue cubes cost 40 cents, and green cubes cost 50 cents, how many different kinds of labels must the third robot be supplied with?

Solution 11

The lowest price is for 3 red cubes: $3 \times 0.30 = 0.90$ dollars. The highest price is for 3 green cubes: $3 \times 0.50 = 1.50$ dollars. The individual prices differ by 10 cents. All prices from 0.90 to 1.50, differing by 10 cents, can be obtained: 0.90, 1.00, 1.10, 1.20, 1.30, 1.40, 1.50.

There are 7 different prices, so the third robot needs 7 different kinds of labels.

Exercise 12

A robot paints identical wooden sticks either green or red. Another robot picks out the painted sticks at random and uses them to make square frames. How many different types of frames can be manufactured in this way?

Solution 12

6 types of frames:

1. all red,
2. all green,
3. 1 green side and 3 red sides,
4. 1 red side and 3 green sides,
5. 2 red sides forming a corner and 2 green sides forming a corner,
6. 2 opposite red sides and 2 opposite green sides.

Exercise 13

For how many two digit numbers is the digit sum equal to the digit product?

Solution 13

Only one number: 22.

Note that the digits cannot be both odd, or one odd and one even. They can only be both even. This is because:

$$odd + odd \neq odd \times odd$$
$$odd + even \neq odd \times even$$

Since the product grows more rapidly than the sum as the digits increase in value and the sum of two even digits cannot be larger than 16, it is easy to rule out any solution other than: $2 + 2 = 2 \times 2$.

Exercise 14

How many + symbols are there in the expression:

$$3 + 3 + 3 + \cdots + 3 = 21300$$

Solution 14

Divide 21300 by 3 to find out how many terms there are on the left side:

$$21300 \div 3 = 7100$$

There are 7100 terms and 7099 + symbols.

Exercise 15

Amira has drawn a flower with 11 petals. She wanted to paint some of the petals yellow. Are there more ways to paint 6 petals or more ways to paint 5 petals yellow?

Solution 15

There would be the same number of possible arrangements in both cases. If Amira paints 5 petals in some pattern, there is an arrangement that corresponds to it exactly if she paints 6 petals:

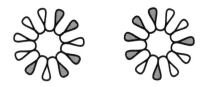

The figure shows there is a one-to-one correspondence between an arrangement with 5 colored petals and an arrangement with 6 colored petals. Therefore, the possible number of arrangements is the same.

Exercise 16

Alfonso is placing boxes of fresh fruit in his store's window. He has 5 boxes of mango and 5 boxes of apples and he wants to arrange them so that no two mango boxes are next to each other. How many different arrangements are there?

Solution 16

Alfonso needs 4 boxes of apples to separate the boxes of mangoes:

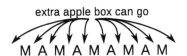

There are now 9 boxes on display, separated by 8 spaces. The remaining apple box can be placed in any of the 8 spaces, plus the two spaces available at the ends of the row.

An arrangement is different from another if they look different from left to right (or from right to left). Therefore, there are 10 possible different arrangements.

Exercise 17

Among her toys, Amira has 5 identical sloths and 5 identical manatees. She wants to arrange them in a circle so that no two sloths are next to each other. How many different arrangements are there?

Solution 17

In the case of a circle, unlike the linear case, there is only one possible arrangement:

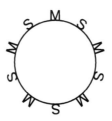

Exercise 18

Four dogs have won prizes in the race organized by Arbax and Lynda: Als, Dux, Los, and Mew. Two dogs tied for the third place and the other two dogs won first and second place. How many possibilities are there for the placements of the four dogs?

Solution 18

There are 12 possibilities. There are 4 choices for the first place winner and 3 choices for second place. The remaining two dogs are automatically in third place.

Exercise 19

Four dogs have won prizes in the race organized by Arbax and Lynda: Als, Dux, Los, and Mew. Two of the dogs have tied for one of the three top places. How many possibilities are there for the placements of the four dogs?

Solution 19

There are three ways of choosing the place where two dogs have tied. For each of these three possibilities, there are 12 different placements of the dogs, as in the previous problem. In total, there are $3 \times 12 = 36$ ways to place the dogs.

Exercise 20

Arbax has three bones: one small, one medium, and one large. He also has three caches: one by the pine tree, one by the shed, and one by the chickencoop. In how many different ways can he store the bones?

Solution 20

Strategy 1: For each bone, there are 3 ways to place it. There are $3 \times 3 \times 3 = 27$ different ways to place the bones.

Strategy 2:

- All three bones are in one cache: there are 3 different ways.

- Two bones are in the same cache and the other is in another cache. There are 3 ways to select the single bone. There are 3 ways to select the cache to place it in. There are 2 choices left for placing the pair of bones. There are $3 \times 3 \times 2 = 18$ different ways.

- Each bone is in a different cache. There are $3! = 6$ ways to place them.

There are $3 + 18 + 6 = 27$ different ways to place the bones.

Solutions to Practice Five

Exercise 1

Answer the following questions about the square table:

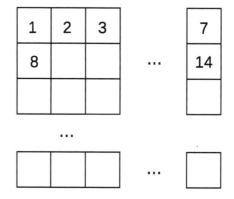

1. How many cells does the table have?
2. Does the table have a cell at its center?
3. What is the number in the bottom right cell?

Solution 1

1. Since it is square, the table has 7 columns, 7 rows, and 49 cells.
2. Since the table has an odd number of rows and columns, it has a cell at its center.
3. The number in the bottom right cell is 49.

Exercise 2

Which number is in the center cell of the square table?

1	3	5
19		

...

17

...

...

Solution 2

The table has $\dfrac{17+1}{2} = 9$ columns and must, therefore, have 9 rows. The center cell must be at the intersection of row 5 and column 5.

The element in row 1, column 5 must be 9.

In the same column, elements in consecutive rows differ by 18. The element in the center cell must be:

$$9 + 18 \times 4 = 81$$

Exercise 3

Is it true that the sum of any 3 numbers we choose from the table below so that no two numbers are on the same column or row, is always the same?

n	n+a	n+2a
n+3a	n+4a	n+5a
n+6a	n+7a	n+8a

Solution 3

By adding three numbers from the table we always obtain a $3 \times n$ term in the sum. Therefore, the answer will be the same if we subtract n from all cells:

0	a	2a
3a	4a	5a
6a	7a	8a

We can also divide the number in each cell by a:

0	1	2
3	4	5
6	7	8

We can write the numbers in each cell like this:

0	1	2
3+0	3+1	3+2
6+0	6+1	6+2

If we choose three numbers from three different columns, we add $0+1+2$ to the sum. Since the three numbers are also on different rows, we add $0+3+6$ to the sum. The sum is $3+9 = 12$ for any such combination of three numbers.

Yes, it is true that the sum is always the same.

Exercise 4

A full set of dominoes has one domino for each possible unique pair of numbers of pips. How many dominoes are there in a full set?

Solution 4

To obtain only unique pairs (not both (2, 3) and (3, 2)), count the pairs so that the first number is smaller than or equal to the second:

$$(0,0)$$
$$(0,1),(1,1)$$
$$(0,2),(1,2),(2,2)$$
$$\cdots$$
$$(0,6),(1,6),(2,6),(3,6),(4,6),(5,6),(6,6)$$

The total number of dominoes is:

$$1 + 2 + 3 + \cdots + 7 = \frac{7 \times 8}{2} = 28$$

Exercise 5

What is the sum of the numbers of pips on the shaded faces of the dominoes? (Dominoes from a single full set have been used.)

Solution 5

Fill the certain values first. After this step, other values will reveal themselves, and so on.

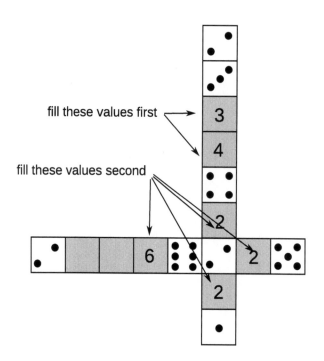

After filling in the above values, we have to start thinking about the values on the faces that are still empty. The leftmost domino has 2 pips on one side and the value on its other side is repeated on the domino that is next in line.

At this point, how many dominoes with 2 pips on one face have we used? We know that there are only 7 such dominoes in a full set. We have used:

$$(5, 2), (4, 2), (1, 2), (6, 2), (3, 2)$$

and only $(0,2)$ and $(2,2)$ are still available. Of these, the leftmost domino cannot be $(2,2)$ since, in that case, the neighboring (*adjacent*) domino would be $(6,2)$ which is no longer available.

The last 2 domino faces have 0 pips.

The sum of the pips on the grey faces is $3 + 4 + 2 + 2 + 2 + 6 = 19$.

Exercise 6

Some numbers are missing from the following magic square. What is their sum?

4		12
		2
8		

Solution 6

Look for rows, columns, or diagonals that differ by only one number.

In this example, denote the numbers that are missing from the main diagonal with a and b:

4		12
	a	2
8		b

From the properties of a magic square, we get that:

$$
\begin{aligned}
4 + a + b &= 12 + a + 8 \\
4 + a + b &= 20 + a \\
4 + b &= 20 \\
b &= 16
\end{aligned}
$$

From here, the magic square is easy to complete because we know that the magic sum is $12 + 2 + 16 = 30$.

4	14	12
18	10	2
8	6	16

The sum of the missing numbers is: $30 + 16 + 18 = 64$

SOLUTIONS TO MISCELLANEOUS PRACTICE

Exercise 1

How many 2-digit prime numbers have one digit that is twice the other?

Solution 1

Since one of the digits is a multiple of 2, it can only be the leftmost digit. Otherwise, the number would be even and therefore not prime.

Let us assume that the last digit is b and the first digit is some even digit that is equal to $2 \times b$. Then, the digit sum of the number is $b + 2 \times b = 3 \times b$. A number with this digit sum is *always* divisible by 3.

There are no numbers with the required property.

Exercise 2

Max, the baker, can make six loaves from a 3 lbs bag of flour with enough flour left over to make another third of a loaf. How many bags of flour does he need in order to make 38 loaves?

Solution 2

If Max uses 3 bags of flour, he can make $3 \times 6 + 1 = 19$ loaves, since he can make 6 loaves out of each bag and one extra loaf out of the leftover flour. Since $38 = 19 \times 2$, Max needs 6 bags of flour to make 38 loaves.

Exercise 3

Amanda earns more money than she needs. Each month, she saves a third of her earnings and spends the rest. She worked for three years, during which her earnings and her expenses remained unchanged. If she stops working and keeps spending at the same rate, for how many months will her savings last?

Solution 3

When she worked, she spent two thirds of her earnings and saved one third. In two months of work, she saved enough money to live for one month without working. Since she worked for 3 years, she can live for 1.5 years on her savings. That is, $12 + 6 = 18$ months.

Exercise 4

A robot produces white square tiles. Another robot produces black square tiles. A third robot picks 4 tiles at random and glues them to make a large square. How many different types of large squares is it possible to make?

Solution 4

There are 6 possibilities:

- All four squares are black.
- All four squares are white.
- 2 white squares on one diagonal and 2 black squares on the other diagonal.
- 2 white squares share a side and 2 black squares share a side.
- 3 white squares and 1 black square.
- 3 black squares and 1 white square.

Note: It is not sufficient to say that there are 2 choices for each square (either black or white) and say that there should be $2 \times 2 \times 2 \times 2 = 16$ different patterns. Among these, some are identical, such as the two examples in the figure:

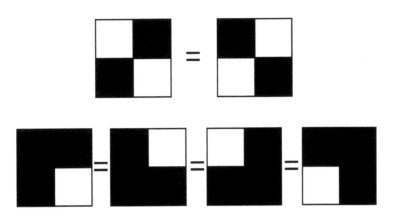

If we carefully account for all these equivalent patterns then the number of possibilities will be reduced to 6, like in the solution.

Exercise 5

Ali and Baba have found a treasure chest filled with gold bars. They cannot use the gold bars to purchase what they need, since one bar is too valuable. They decided to cut the bars into coins. From each bar they can make 4 coins with some gold left over. The leftover gold from 6 bars can be used to make one whole bar. How many bars did they find if they were able to make 172 coins?

Solution 5

If Ali and Baba turn 6 bars into coins, they obtain $6 \times 4 = 24$ coins and 6 leftovers which can be made into a new bar. From this bar, they can make 4 more coins, for a total of 28 coins and still have one sixth of a bar left over.

Since $172 = 6 \times 28 + 4$, they must have had 36 bars. Out of each group of 6 bars they made 28 coins and for each group there was a sixth of a bar left over. Since there are 6 groups of 6 bars, one more bar can be made out of the leftover from each group. From this bar, they can make 4 more coins and be left with another sixth of a bar.

The following table helps keep track of the data:

Bars	Coins	Leftover
6	24	$6 \times \frac{1}{6} = 1$
6	28	$\frac{1}{6}$
36	168	1
36	172	$\frac{1}{6}$

Exercise 6

From the pairs of sets defined below, select the pair that represents disjoint sets:

(A) the set of people on Dina's street who live at even numbered houses and the set of people who own dogs;

(B) the set of days with rain and the set of days with sun in Lila's town during August 2012;

(C) the set of mothers with only two children and the set of mothers with only three children on Amira's street;

(D) the set of houses with solar panels and the set of houses with fruit trees on Dina's street.

Solution 6

The answer is (C).

(A) The set of people on Dina's street who live at even numbered houses and the set of people who own dogs are not disjoint since there may be people who live at an even number house and own a dog at the same time.

92

(B) The set of days with rain and the set of days with sun in Lila's town during August 2012 are not disjoint, since it is possible for there to be both sun and rain on the same day.

(C) The set of mothers with only two children and the set of mothers with only three children on Amira's street are disjoint. A mother can have either two or three children.

(D) The set of houses with solar panels and the set of houses with fruit trees on Dina's street are not disjoint since it is possible for a house to have both fruit trees and solar panels.

Exercise 7

Amanda is choreographing a new dance. In it, twelve dancers stand in a row, equally spaced, waving scarves. Amanda then dances from one end of the row to the other end, passing in front of each dancer. It takes Amanda 4 seconds to dance from the first dancer to the fifth dancer. How many seconds does it take her to dance from the one end of the row to the other end of the row?

Solution 7

There are four gaps between the five dancers. In 1 second, Amanda dances across one gap. There are 11 gaps between 12 dancers. It will take Amanda 11 seconds to dance across 11 gaps.

Exercise 8

What is the last digit of the product:

$$3 \times 33 \times 333 \times 3333 \times \cdots \times \underbrace{33\cdots3}_{111 \text{ digits}}$$

Solution 8

It is the same last digit as that of the product:

$$\underbrace{3 \times 3 \times 3 \times \cdots \times 3}_{111 \text{ times}}$$

Since the last digit of a repeated product of 3s forms the sequence:

$$3, \; 9, \; 7, \; 1, \; 3, \; 9, \; \ldots$$

we have to divide 111 by 4:

$$111 = 4 \times 27 + 3$$

The product will end in 7.

Exercise 9

Of the following pairs of sets, which ones represent a set that is included in the other? Check all that apply.

(A) the set of dogs who won the race and the set of dogs who trained for the race;

(B) the set of prime numbers and the set of fractions;

(C) the set of vegetables and the set of boxes in Alfonso's store;

(D) the set of cars on the road and the set of passengers in those cars.

Solution 9

Only choice (B) represents a set and a subset.

(A) The race may have been won by a dog that did not train for the race.

(B) Like any integer, a prime number can be written as a fraction. For example, $41 = \dfrac{41}{1}$. Therefore, prime numbers are fractions. The set of prime numbers is a subset of the set of fractions.

(C) Although the vegetables are inside boxes, this does not mean that the sets are inclusive. For a set to be included in another set, its elements have to be the elements of the larger set as well. Vegetables and boxes form two disjoint sets, since no vegetable can be a box.

(D) Although the passengers are inside the cars, this does not mean that the sets are inclusive. No passenger can be a car, therefore the sets must be disjoint.

Exercise 10

How many different ways of painting a square black are there in a 4×4 grid of white squares?

Solution 10

By rotating the grid, we obtain equivalent figures such as the following:

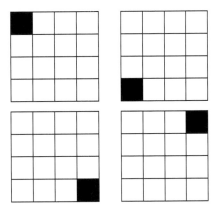

This means that only the squares in one quarter of the grid produce distinct colorings. There are 4 distinct colorings.

Note: the grid is assumed to be drawn on one side of a paper sheet, which is why the grids below are not considered equivalent:

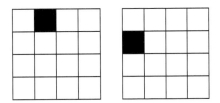

Exercise 11

Enumerate the elements of each set:

1. the set of prime numbers smaller than 20;

2. the set of planets in the solar system;

3. the set of (possibly meaningless) words that can be formed with the letters M, E, and W;

4. the set of playing cards that are multiples of 5.

Solution 11

(A) {2, 3, 5, 7, 11, 13, 17, 19}

(B) {Mercury, Venus, Earth, Mars, Jupiter, Saturn, Uranus, Neptune}

(C) {MEW, MWE, WEM, WME, EWM, EMW}

(D) {5♢, 5♡, 5♣, 5♠, 10♢, 10♡, 10♣, 10♠}

Exercise 12

How many elements does each set have?

1. the set of factors of 441;

2. the set of students from the 4th in line to the 24th in line;

3. the set of digits used to write integer numbers from 1 to 99;

4. the set of 5 digit numbers with a digit sum of 4 and a non-zero digit product.

Solution 12

1. Factor 441 into primes:

```
441  │ 3
147  │ 3
 49  │ 7
  7  │ 7
  1  │
─────┼────
441 = 3 x 3 x 7 x 7
```

The set of divisors is: $\{1, \ 3, \ 7, \ 9, \ 21, \ 49, \ 63, \ 147, \ 441\}$ and it has 9 elements.

2. There are: $24 - 4 + 1 = 21$ students. The set has 21 elements.

3. 10 digits are used to write the numbers. The set has 10 elements.

4. There are no 5 digit numbers with a digit sum of 4 and a non-zero digit product. A digit sum of 4 can be achieved with 4 digits of 1, the remaining digit must be zero and will make the product zero. If we use larger digits, there are even more digits of zero. The solution set is *the empty set* - it has zero elements.

Exercise 13

How many rectangles with integer sides have a perimeter of 14?

Solution 13

Let the width of the rectangle be W and the length be L. Then, the perimeter is equal to:

$$L + L + W + W = 2 \times (L + W)$$

If the perimeter is 14 then:

$$2 \times (L + W) = 14$$
$$L + W = 7$$

L and W are positive integers. This is a Diophantine equation. There are 3 distinct solutions:

$$
\begin{aligned}
L &= 1, \ W = 6 \\
L &= 2, \ W = 5 \\
L &= 3, \ W = 4
\end{aligned}
$$

Rectangles obtained by interchanging L with W are not different from these, but are just their rotations.

Exercise 14

What are the last 3 digits of the product:

$$20! = 1 \times 2 \times 3 \times \cdots \times 20$$

Solution 14

Since there are 4 multiples of 5 in the product and sufficient factors of 2, we can form four 2×5 factor pairs. Each of these pairs adds a zero at the end of the final result. The product ends in 4 zeros. The last 3 digits are 000.

Exercise 15

Amira has lots of magnetic balls and rods. She made the following pattern:

How many balls and how many rods did she use?

Solution 15

There are 6 rods and 6 balls in each hexagon and 9 connecting rods between hexagons. Amira used: 60 balls and 69 rods.

Exercise 16

In which of the following rectangular grids is it possible to color a third of the small squares black?

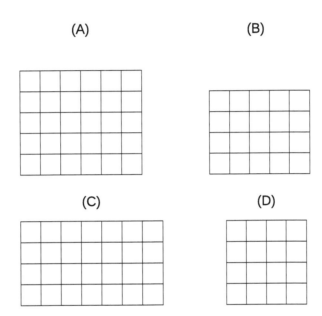

Solution 16

Only grids with a number of cells that is a multiple of 3 will work. The answer is (A).

Exercise 17

How many digits of 0 are there in the result of the operation:

$$\underbrace{444\cdots44}_{444 \text{ digits}} \times 7$$

Solution 17

Apply the multiplication algorithm. Because 7×4 has a carryover, the result of the multiplication will have 445 digits:

$$\underbrace{444\cdots44}_{444 \text{ digits}} \times 7 = 3\underbrace{000\cdots0}_{443 \text{ digits}}8$$

There are 443 digits of zero.

Exercise 18

Lila's math club walked to the school's parking lot for an experiment. Lila noticed that there were 28 blue cars in the parking lot. Of these, 15 had tinted windows and 19 had roofracks. What could have been the largest possible number of blue cars with tinted windows but without roofracks?

Solution 18

Since $15 + 29 = 34$ and 34 exceeds 28 by 6, 6 blue cars had tinted windows and roofracks. At most $15 - 6 = 9$ cars had tinted windows and no roofracks.

Exercise 19

How many 2-digit numbers are divisible by their tens digit?

Solution 19

- All 10 numbers from 10 to 19

- 5 numbers from 20 to 29 (all the even numbers);

- 4 numbers from 30 to 39 (all multiples of 3);

- 3 numbers from 40 to 49 (all multiples of 4);

- 2 numbers from 50 to 59 (50 and 55);

- 2 numbers from 60 to 69 (60 and 66);

- 2 numbers from 70 to 79 (70 and 77);

- 2 numbers from 80 to 89 (80 and 88);

- 2 numbers from 90 to 99 (90 and 99);

There are 32 numbers with the required property.

Exercise 20

Amira had some red and some blue blocks in a box. If Lila takes some blocks out of the box blindfolded, she has to take out at least 11 blocks to be sure she has at least 2 blue blocks and she has to take out at least 8 blocks to be sure she has at least 3 red blocks. How many blocks did Amira have in the box?

Solution 20

If Lila has to take out at least 11 blocks to be sure she has at least 2 blue blocks, then there must be $11 - 2 = 9$ red blocks in the box.

If Lila has to take out at least 8 blocks to be sure she has at least 3 red blocks, then there must be $8 - 3 = 5$ blue blocks in the box.

Amira's box contains 14 blocks.

Exercise 21

Which of the following arrangements are possible using a full set of dominoes?

Solution 21

Only (D) is possible. (A) through (C) are not possible because there is only one domino of each kind in a set.

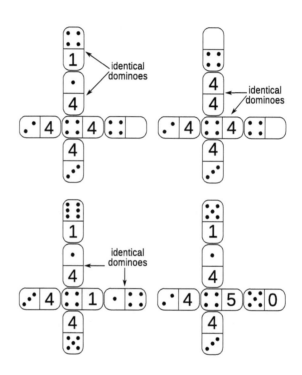

Exercise 22

Dina and Lila have 2 red chickens, one white chicken, and one black chicken. When the sun goes down, the chickens retire to their nests where they sit in a line. In how many different ways can they sit if the red chickens always sit side by side?

Solution 22

Remove one of the two red chickens. There are $3! = 6$ possible arrangements in this case. The two red chickens, however, are two different animals. Wherever they are placed it is possible to switch them around and obtain two different arrangements. For this reason, there are $6 \times 2 = 12$ possible arrangements.

Exercise 23

Alfonso has to place some egg cartons in the store's window for advertisement. He has two boxes labeled "cage free," one box labeled "free range," and one box labeled "happy hens." In how many ways can he make a row of boxes out of them if the two "cage free" boxes have to be placed side by side?

Solution 23

Remove one of the two "cage free" boxes. There are $3! = 6$ possible arrangements in this case. Since the two boxes are identical, we do not obtain different arrangements by switching them around. The answer is 6. Notice the difference from the previous problem, where the two chickens, though both red, are not identical. Dina and Lila can tell which one is which.

Exercise 24

The three machines in the figure can be combined to produce an output when a number is input. How many different results can we obtain from a single input if we are allowed to use each machine once but in any order?

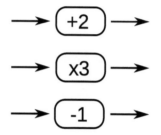

Solution 24

We can generate 4 different answers. Let us use In as an input number:

$$
\begin{aligned}
\text{Out} &= (\text{In} + 2 - 1) \times 3 = 3 \times \text{In} + 3 \\
\text{Out} &= \text{In} \times 3 + 2 - 1 = 3 \times \text{In} + 1 \\
\text{Out} &= (\text{In} + 2) \times 3 - 1 = 3 \times \text{In} + 5 \\
\text{Out} &= (\text{In} - 1) \times 3 + 2 = 3 \times \text{In} - 1
\end{aligned}
$$

Exercise 25

How many rectangles are there in the figure?

Solution 25

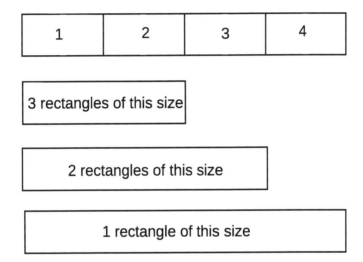

There are $4 + 3 + 2 + 1 = \dfrac{4 \times 5}{2} = 10$ rectangles in total.

Exercise 26

How many rectangles are there in the figure?

1	2		20

Note: The number of rectangles is a triangular number.

Solution 26

Generalize the solution of the previous problem. There are 210 rectangles in total:

$$1 + 2 + 3 + \cdots + 20 = \frac{20 \times 21}{2} = 21 \times 10 = 210$$

Exercise 27

How many rectangles are there in the figure?

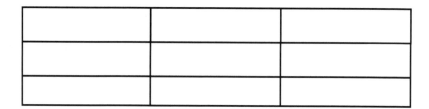

Solution 27

There are $3+2+1 = 6$ rectangles in each of the 3 rows and $3+2+1 = 6$ rectangles in each column. The total number of rectangles is $6 \times 6 = 36$.

Exercise 28

In how many different ways can we place the black rectangle inside the white rectangle so that it covers two squares exactly?

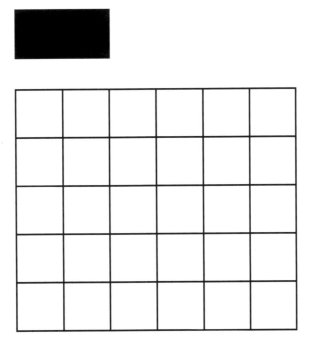

Solution 28

The white rectangle is a grid with 6 horizontal and 5 vertical cells.

If the black rectangle is horizontal, we can place it in 5 different positions on each row. In total $6 \times 5 = 30$ different positions.

If the black rectangle is vertical, we can place it in 4 different positions in each column. In total $4 \times 6 = 24$ different positions.

The total number of different positions is $30 + 24 = 54$.

Competitive Mathematics Series for Gifted Students

Practice Counting (ages 7 to 9)
Practice Logic and Observation (ages 7 to 9)
Practice Arithmetic (ages 7 to 9)
Practice Operations (ages 7 to 9)

Practice Word Problems (ages 9 to 11)
Practice Combinatorics (ages 9 to 11)
Practice Arithmetic(ages 9 to 11)
Practice Operations (ages 9 to 11)

Practice Word Problems (ages 11 to 13)
Practice Combinatorics (ages 11 to 13)
Practice Arithmetic and Number Theory (ages 11 to 13)
Practice Algebra and Operations (ages 11 to 13)
Practice Geometry (ages 11 to 13)

Practice Word Problems (ages 12 to 15)
Practice Algebra and Operations (ages 12 to 15)
Practice Geometry (ages 12 to 15)
Practice Number Theory (ages 12 to 15)
Practice Combinatorics and Probability (ages 12 to 15)

This is a series of practice books. With the exception of a few reminders, there are no theoretical explanations. For lessons, please see the resources indicated below:

Find a set of free lessons in competitive mathematics at www.mathinee.com. Addressing grades 5 through 11, the *Math Essentials* on www.mathinee.com present important concepts in a clear and concise manner and provide tips on their application. The site also hosts over 400 original problems with full solutions for various levels. Selectors enable the user to sort essentials and problems by test or contest targeted as well as by topic and by the earliest grade level they can be used for.

Online problem solving seminars are available at www.goodsofthemind.com. If you found this booklet useful, you will enjoy the live problem solving seminars.

For supplementary assessment material, look up our problem books in test format. The "Practice Tests in Math Kangaroo Style" are fun to use and have a well organized workflow.

8987974R00062

Printed in Great Britain
by Amazon.co.uk, Ltd.,
Marston Gate.